Astronomy in Zyn XYZ

© 2024 Diogo de Souza
All Rights Reserved.

Contact Information:
diogodesouza7@gmail.com
diogodesouza7@hotmail.com

Introduction:

This book is a collection of Photos of the sky at night taken with a short time exposure (10 seconds), using a DSLR Camera, and a Telescope mounted on a Motorized Tripod. All of the images were taken under heavily light polluted skies of the City of Dallas Fort Worth, but made possible due to the tracking device and image processing using an Android Phone. It is not impossible to perform stargazing in the middle of a large Metropolis, but it is always a good idea to have a trip outside of the City to stargaze. Unfortunately, the equipment is rather large, and I was not able to venture beyond the light polluted regions and decided to see what results would be gotten by just shooting in the direction of the Stars from a place that was more convenient such as the Backyard of my Home. The best of all the results is summarized in this book with images of the most common targets in the Night Sky as seen from North America. Each image has a story to tell, and I expand the wonders of these colorful works of Art in the Night Sky into Philosophy and Scientific Information about the objects shown. The Night Sky is truly beautiful, but it can only be enjoyed from a Large City with good equipment that is capable to reduce the negative effects of light pollution.

Table of Contents:	**Page:**
Moon and Planets	4
Star Clusters	12
Nebulas and Galaxies	67
Eclipses	85
Tables	91

Moon and Planets

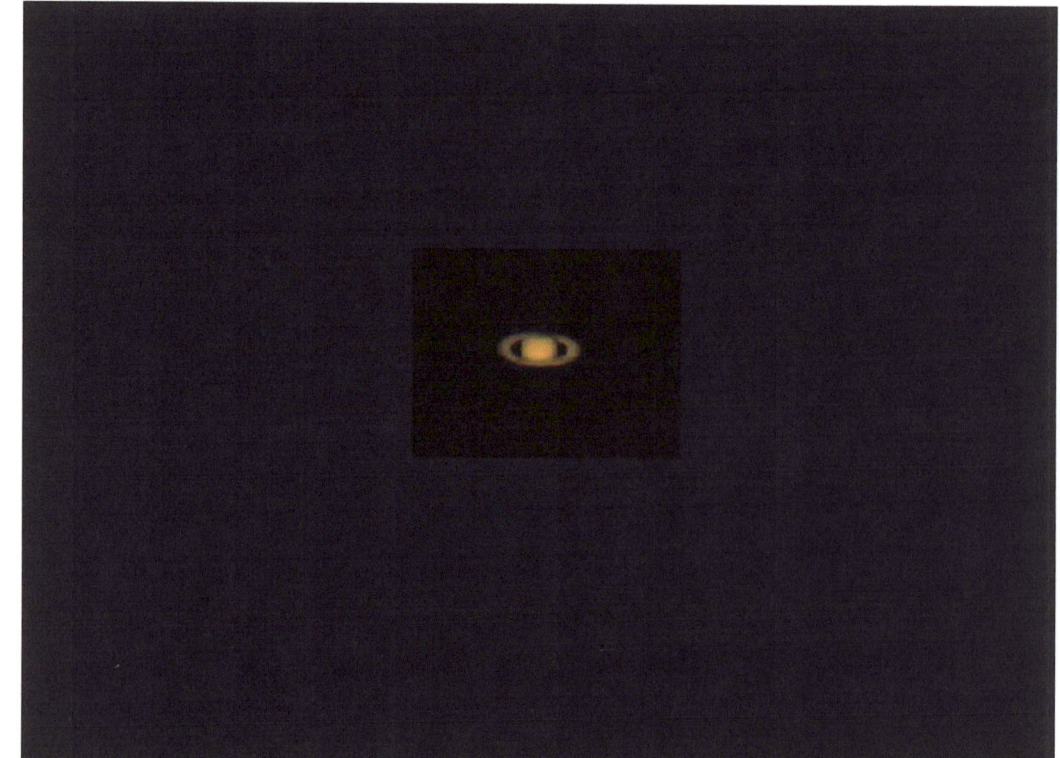

The Moon

The Moon

The Moon is Earth's Natural Satellite and has been important not only in the formation and History of our Planet, but its motion was used throughout centuries by humanity in order to measure Time, create Calendars, and has always deeply influenced Human Culture and Myths. It is a sphere of rock, metals and minerals, orbiting our Planet. It has phases and has an orbital Period of about a Month.

Moon and Venus

Venus is the Morning Star, but it can also be called the Evening Star since it is closer to the Sun than the Earth and it appears always next to our Sun in the early Morning and early Evening. Here in the image above, Venus is to the left of the Crescent Moon as seen through Binoculars. The sky is filled with signs and many of these Celestial Encounters led in the past to Myths and superstition. An encounter of Venus with the Moon could for the Ancient People mean something on Earth, but obviously these are just Celestial Positions in the sky for us to wonder and that may hold meanings only in our hearts.

Planet Uranus

Uranus looks like a faint ghostly blueish white globe in the sky. It is beautiful since it appears to be clearly not a Star, and it has a unique ghostly aspect that makes it a Planet worth a Time to contemplate.

Planet Jupiter

When looking at the night sky, a very bright Star that appears to move against the background Constellations over Time is not a Star but rather a Planet. Jupiter appears like a very bright point of light and when seen with a Telescope, four Moons can be seen around it in orbit. A closer look at the Planet reveals bands in its Atmosphere and a Planet that is not perfectly circular. It seems to bulge towards the Equator due the Centrifugal Forces and is seen tilted on its Axis. Jupiter is the King of the Night Sky for its brightness and is fascinating.

Planet Saturn

First observed from my Telescope in the year 2003, Saturn was the first Planet that I observed and is one of the reasons of what led me into Astronomy. Its rings and yellow color are a unique feature in the Night Sky. It is not the only Planet in the Solar System with rings, but it is the only one that can be seen with rings using simple equipment such as available to most Amateur Astronomers. Its rings can change size appearance as the Planet orbits the Sun due to the fact that the Planet is tilted on its axis and different angles lead to different viewpoints of the rings. In the picture above the rings appear thin and edge on.

Planet Saturn

In this image Saturn appears closer, and the angle of its tilt favors a better view of its rings. The slightly darker portion of Saturn near its top is the shadow of the Planetary Globe over that portion of its rings. Saturn is very yellow and can be spotted with any simple Telescope. It appears much fainter in the Night Sky than Jupiter but is easily visible with rings. Through Binoculars it appears as a yellow oval object, but through Telescope it is a reason for much contemplation and the sight is unbelievable.

Star Clusters

Center of Northern Cross

Perseus Double Cluster

This Cluster is favorite among Astronomers for being easily visualized even from small Telescopes. It is such a dense cluster that it almost looks like two Globular Clusters next to each other. They were formed from Nebulas and are part of the same Nebula System.

Sagittarius Star Cloud

Near the direction of the Center of the Milky Way Galaxy, a dense number of Stars is seen in the Northern Portion of Sagittarius as several thousand clustered with Stars brighter than others. As we look towards the Center of the Galaxy, more Stars and Dust can be seen.

Star Vega

One of the brightest Stars seen from the Northern Hemisphere, Vega can be clearly seen in the Constellation Lyra even from Light Polluted Skies. Under heavy Light Pollution, this Star can often times be the only Star easily visible in the Night Sky. With Telescopes and Binoculars, Vega is a show and in the image above it even appears to have spikes from the diffraction of its light in the Telescope Lens.

Star Sirius

This is the brightest Star seen from the Northern Hemisphere. It is located in the Northern Part of the Constellation Canis Major and it can be clearly seen even under light polluted skies of big cities. When looking at the Star from a Telescope's Eyepiece it appears to twinkle in the colors blue, red, and white. It is located in a region of the sky right along the Milky Way where there are a great number of Stars. Sirius has been part of many Mythologies of past Civilizations becoming a sky wonder for many people throughout many ages.

Star Sirius

Sirius glows brightly in the sky and it is also not very far from our Solar System being a close neighbor only within 8 Light Years away. It is actually a Binary Star although it appears as single Star as seen from a Small Telescope. Sirius is also known as the Dog Star due to it being the brightest Star at Constellation Canis Major. It is best seen in the sky during the winter in Northern Hemisphere and it is located south of the Constellation Orion.

Bright Star is Alnilam in the Center of Orion Belt.

Bright Star is Alnitak in Eastern Portion of the Orion Belt.

Bright Star is Mintaka in the Western Portion of the Orion Belt.

Pleiades

The Pleiades appear as a faint smudge in the Night Sky. With Telescopes and Binoculars many Stars are seen and there appears to be some dust in between these points of light making its view spectacular.

Pleiades

Perseus Double Cluster

Better equipment and using different image filters reveal even more Stars in this Double Cluster. These are all new Stars recently formed from Interstellar Gas that permeates all of Space. Stars even appear to differentiate in their colors, and some are brighter than others. The image provides a Three-Dimensional look into this portion of the Cosmos.

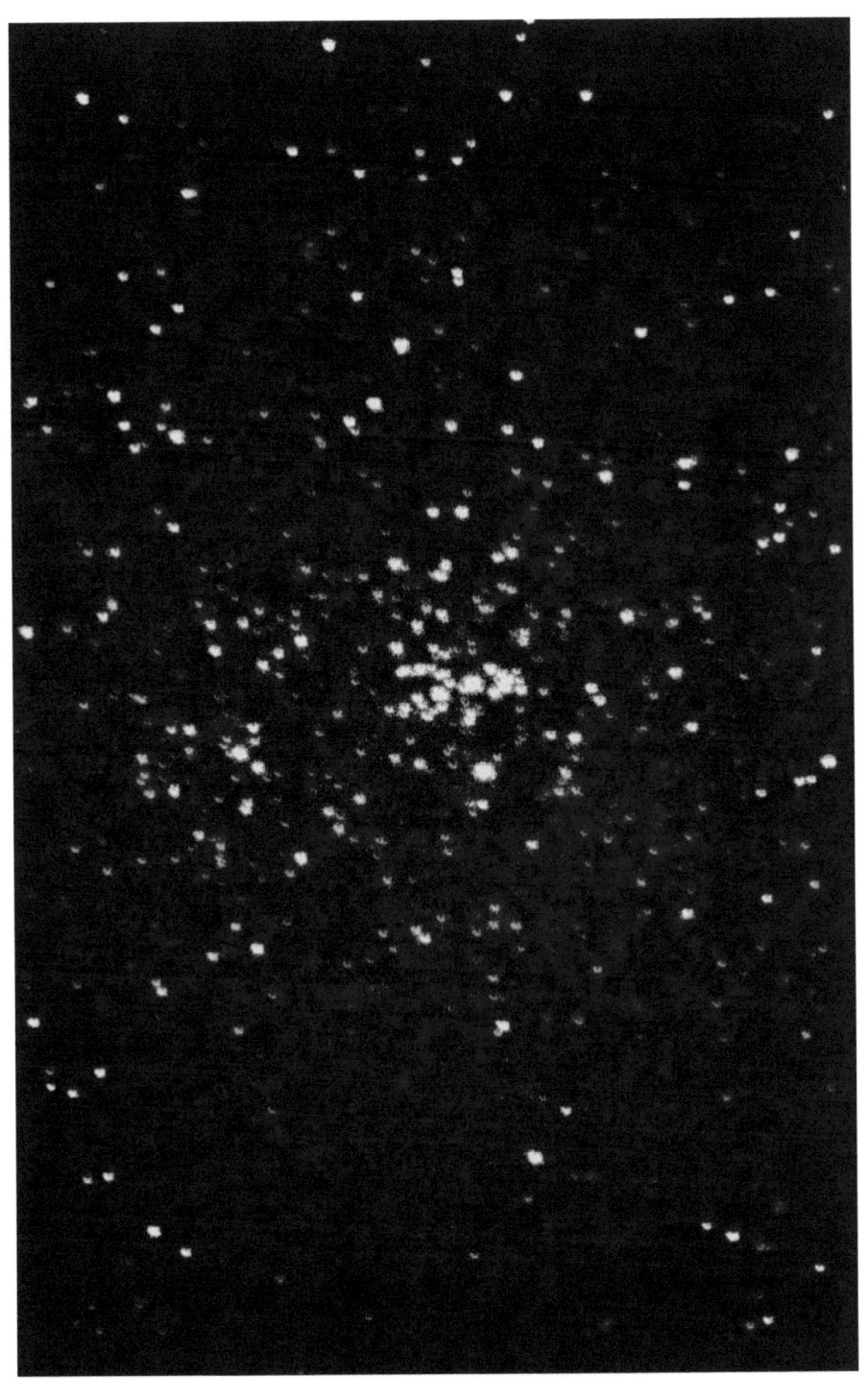

Perseus Double Cluster

Albireo

Bright Double Star Albireo in the background of Stars in the Center of the Summer Triangle.

Albireo Double Star

Albireo is a bright Double Star in the center of the Summer Triangle. The bigger Star is red, and the smaller Star is blue. It is a very popular Binary Star since it can be easily found when moving the eyes to the east of the Constellation Lyra. Any small Telescope reveals it as a Binary System.

145 Canis Majoris Double Star

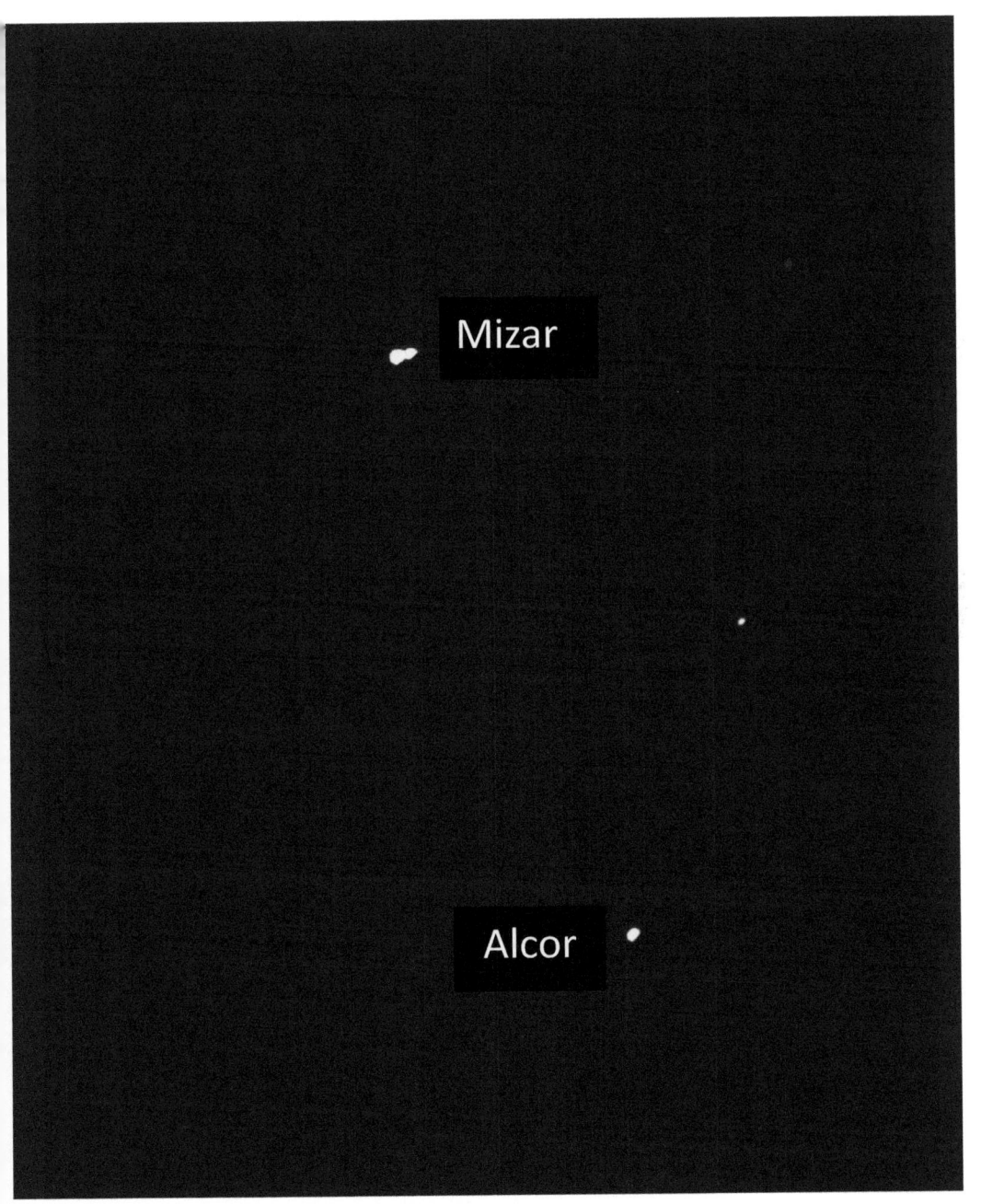

Mizar Double Star

Cor Caroli Double Star

Binary Stars can be seen in all direction being the most common type of Star in the Galaxy.

Cluster at the Center of Northern Cross

C 50

Cluster of Bright Stars in the Center of a Large Nebula not visible in the image above.

Little Beehive

Brightest Star Cluster in the Constellation Canis Major South of the Equator.

M41 (Little Beehive Cluster)

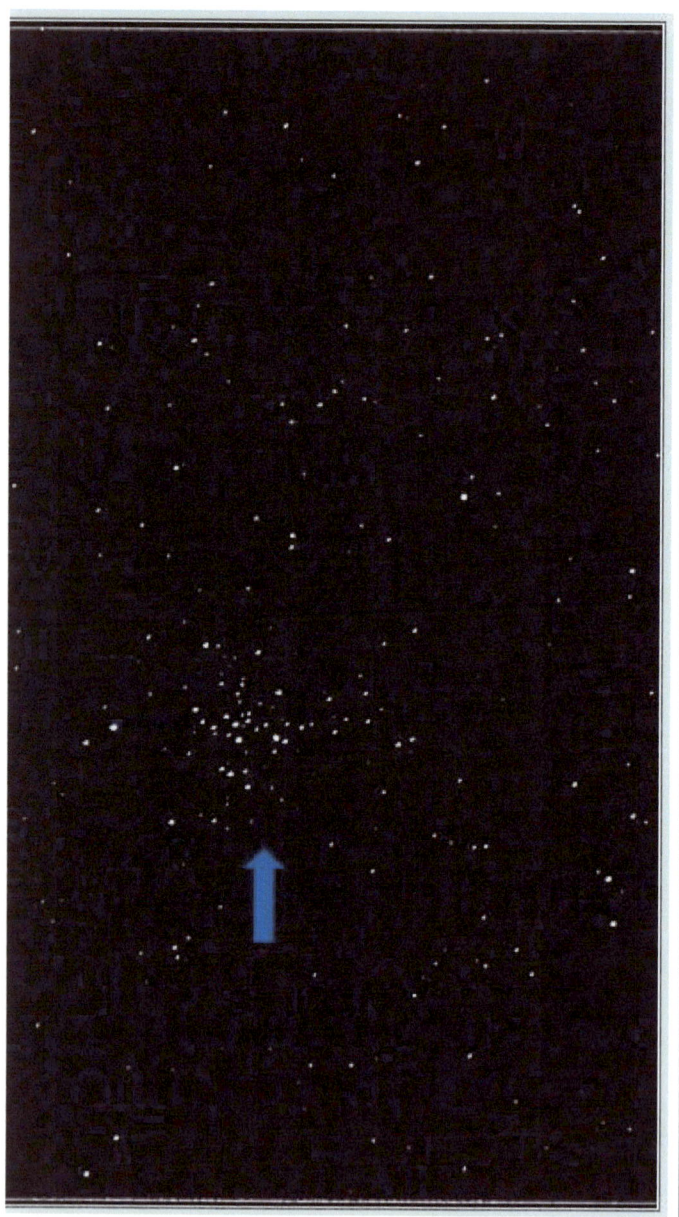

South of The Bright Star Sirius is the Little Beehive Cluster. It is called that way for its resemblance of the Beehive Cluster in the Constellation Cancer. Although smaller in the Night Sky it is bright and appears to be composed of many more Stars compressed in a smaller region of Space.

M41 Little Beehive

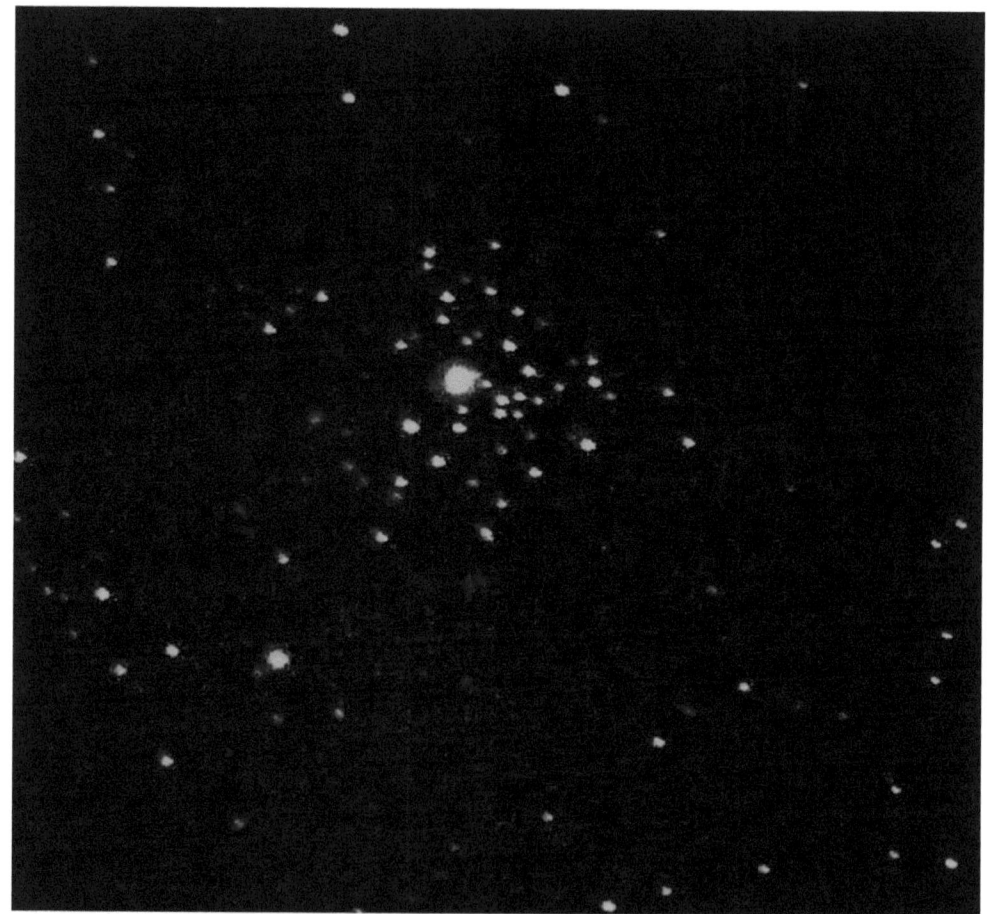

C 64

The image above is of a very interesting open cluster in the Southern Portion of the Constellation Canis Major. It is unique in the fact that it is shaped like a shield with a bright Star at its center.

C 64 seen from far away clearly resembles a Star Cluster. It is located in a region of the Sky with a dense population of Stars right along the Milky Way Band.

Hercules Globular Cluster

This is the most popular Globular Cluster viewed from the Northern Hemisphere also because of how easy it is to find the Night Sky. It is located in the Western Portion of the Constellation Hercules between two bright Stars.

Hercules Globular Cluster

It contains several hundreds of Stars and possibly a Black Hole is located at its center. It was the target of the Arecibo's Message from the Radio Telescope Observatory in Puerto Rico who sent a signal in 1974 containing encoded information about our World and our Civilization. The message was sent hoping on a possible contact with an Extraterrestrial Civilization along the way that is capable of Radio Communication.

M 4 Globular Cluster

A very faint Globular Cluster in the Constellation Scorpius seen from the Northern Hemisphere. It is in a position close to the horizon in the southern direction, where the effect of Light Pollution is intense. It is practically invisible, and it can only be seen with a DSLR Camera attached to a Telescope. Astronomy is a very fun Science, but it requires equipment to be fully enjoyed. M4 is an example of a cluster that is almost invisible but when viewed outside of Light Pollution it is more evident. This shows how much humanity has lost since the advent of electricity.

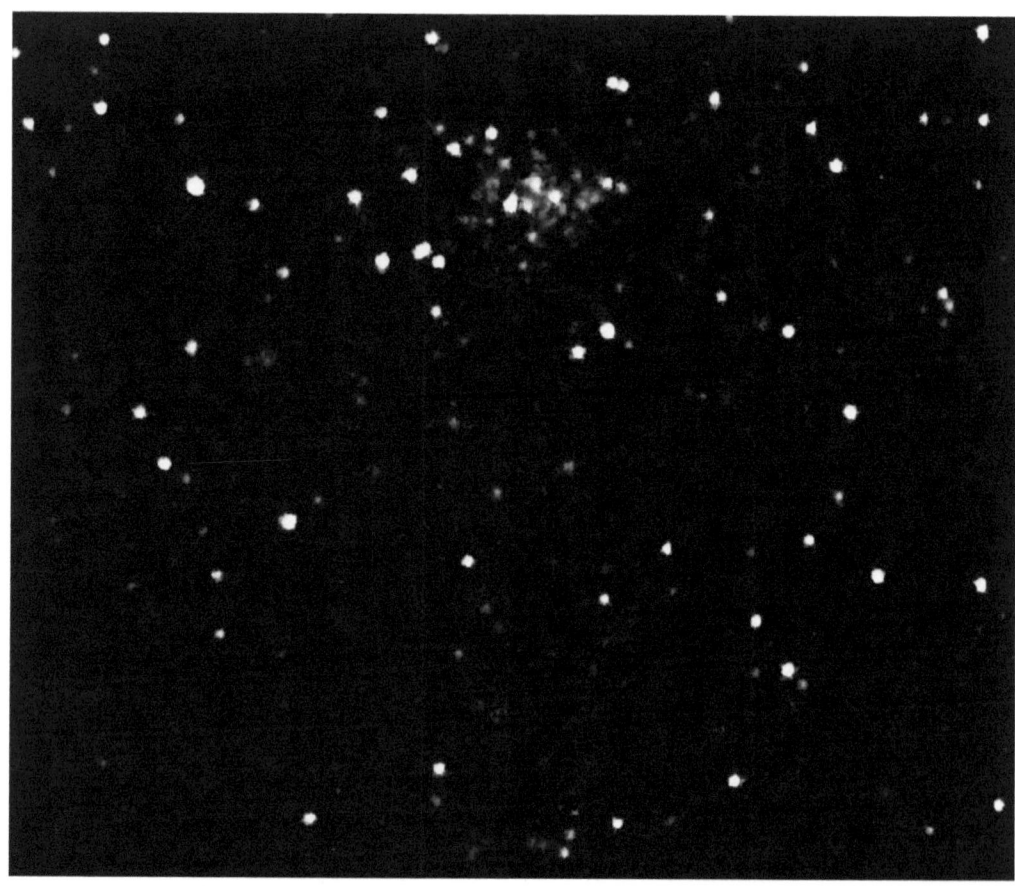

M 71 Globular Cluster

This is another cluster that is not possible to be seen when placing a human eye in the Telescope's Eyepiece. The image above is what a DSLR Camera attached to the Telescope was able to capture. It is basically what would be possible to see without any Light Pollution, but a DSLR Camera helps locate objects washed out by the glare of electricity.

M3 Globular Cluster

This cluster appears very compact and dense located West of the Constellation Boötes. It is a very bright cluster capable to be seen with Binoculars even in a Large City. Its Globular Shape is very well defined more so than other Globular Clusters. It also very easy to find in the Night Sky when gazing a bit West and North from the red Star Arcturus. It is not located on the band of the Milky Way but rather in the direction outside proving that Globular Clusters are found everywhere around and outside of the Galactic Disk.

M 12 Globular Cluster

Globular Cluster located inside the Constellation Ophiuchus North of Globular Cluster M 10. It appears somewhat fainter, but many Stars can be distinguished in the image.

M 10 Globular Cluster

Globular Cluster located inside the Constellation Ophiuchus South of the Globular Cluster M 12. Together they make two great targets for a Constellation that is often overlooked and that could very well had been the 13th Constellation of the Zodiac but never defined that way,

M 22 Globular Cluster

This is truly the largest Globular Cluster that I was able to capture with my camera. Many Stars can be distinguished and it is a show of lights in the Northern Region of the Constellation Sagittarius. Around the Cluster other Stars are visible in a dense region of the Night Sky close to the direction of the center of the Mily Way Galaxy.

M 5 Globular Cluster seen from far away at the tip of the arrow. It is located on the western portion of the Constellation Serpens and is a great target East of the Star Arcturus adorning the sky.

Cassiopeia, Perseus, and Pleiades

 Orion

This is a great image of the Southern Portion of the Constellation Orion. There are many Stars in this region and the Orion Nebula appears very bright in the center while the three Stars of the Orion Belt are clearly seen in a diagonal line.

Beehive Cluster

This Cluster is very sparse located at the very center of the Constellation of Cancer and can be seen with any Binoculars and Telescope. It is clearly a cluster but the fact of it being sparse leads to a low density of Stars. Possibly a Cluster that is expanding from a point that once in the past was a Nebula. All Clusters first began as a Nebula, then as a Cluster that expands, and the end result is no more Cluster with Stars spread in the Galaxy.

M47

M47

M46

M46

The yellow arrow points to a faint Planetary Nebula called NGC 2438 on the edge of M46.

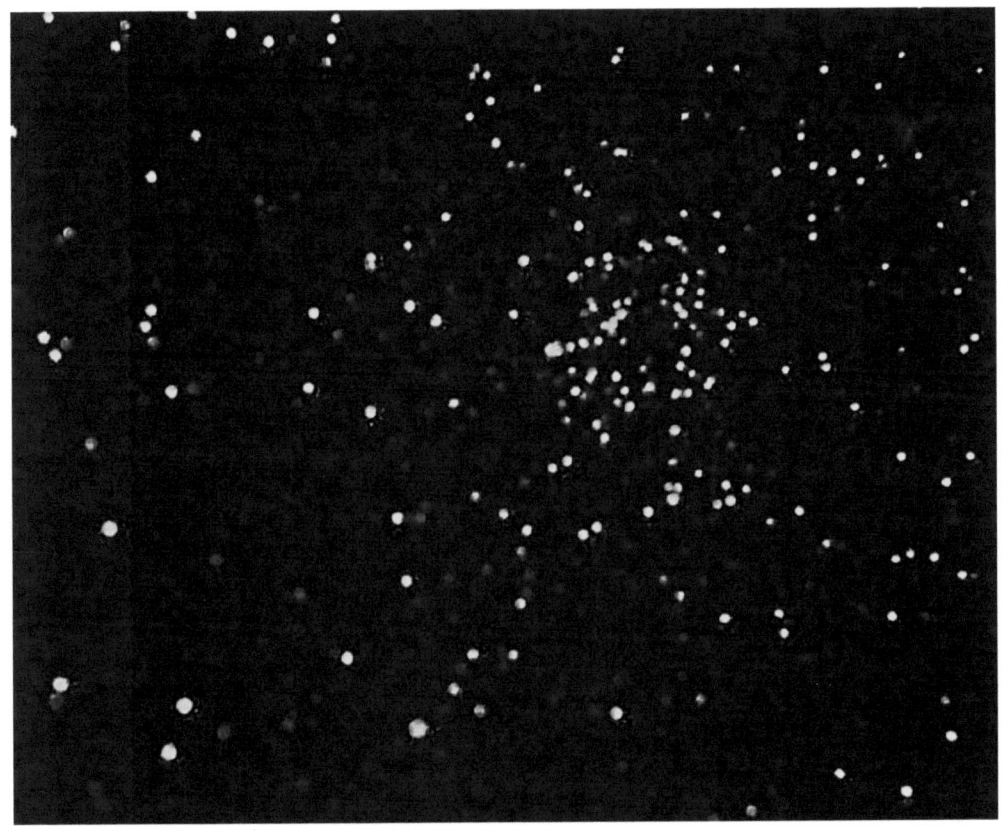

M35

M 35 is a beautiful Cluster adorning the Western Portion of the legs of the Constellation Gemini. It has a sort of oval shape in the form of a Cluster that appears to be unfolding and spreading in Space. Some Stars look aligned and departing from the Cluster that is expanding. It is located along the band of the Milky Way Galaxy and is best viewed during the Winter in the Northern Hemisphere.

M 35

Here again is a nice and sharp view of M 35 which is a Cluster that is about 2,970 Light Years away from us.

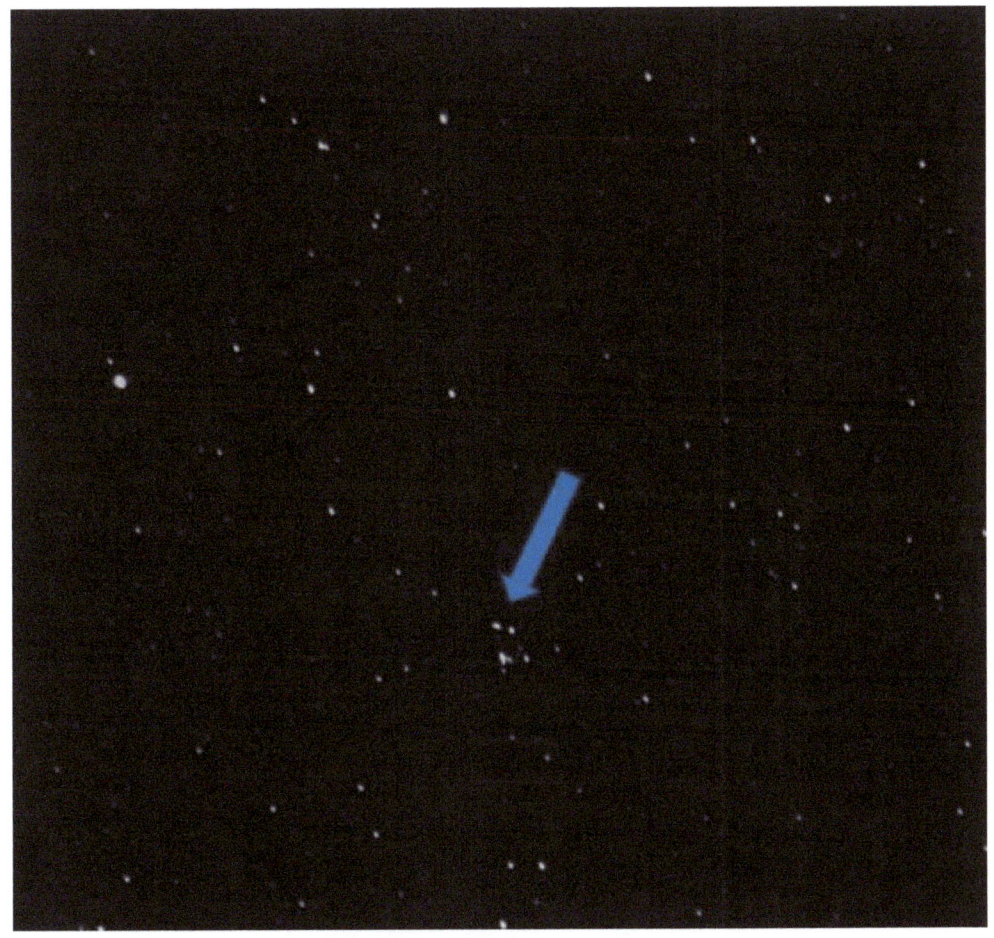

Cluster 37 at tip of arrow

This Cluster is also called NGC 2169 and is located in the Constellation Orion at most precisely at Orion's Eastern Arm. When seen with more magnification the orientation of the Stars resembles the numbers 3 and 7 thus its nickname. When seen from far away like in the picture above, it is a small gathering of Stars close together.

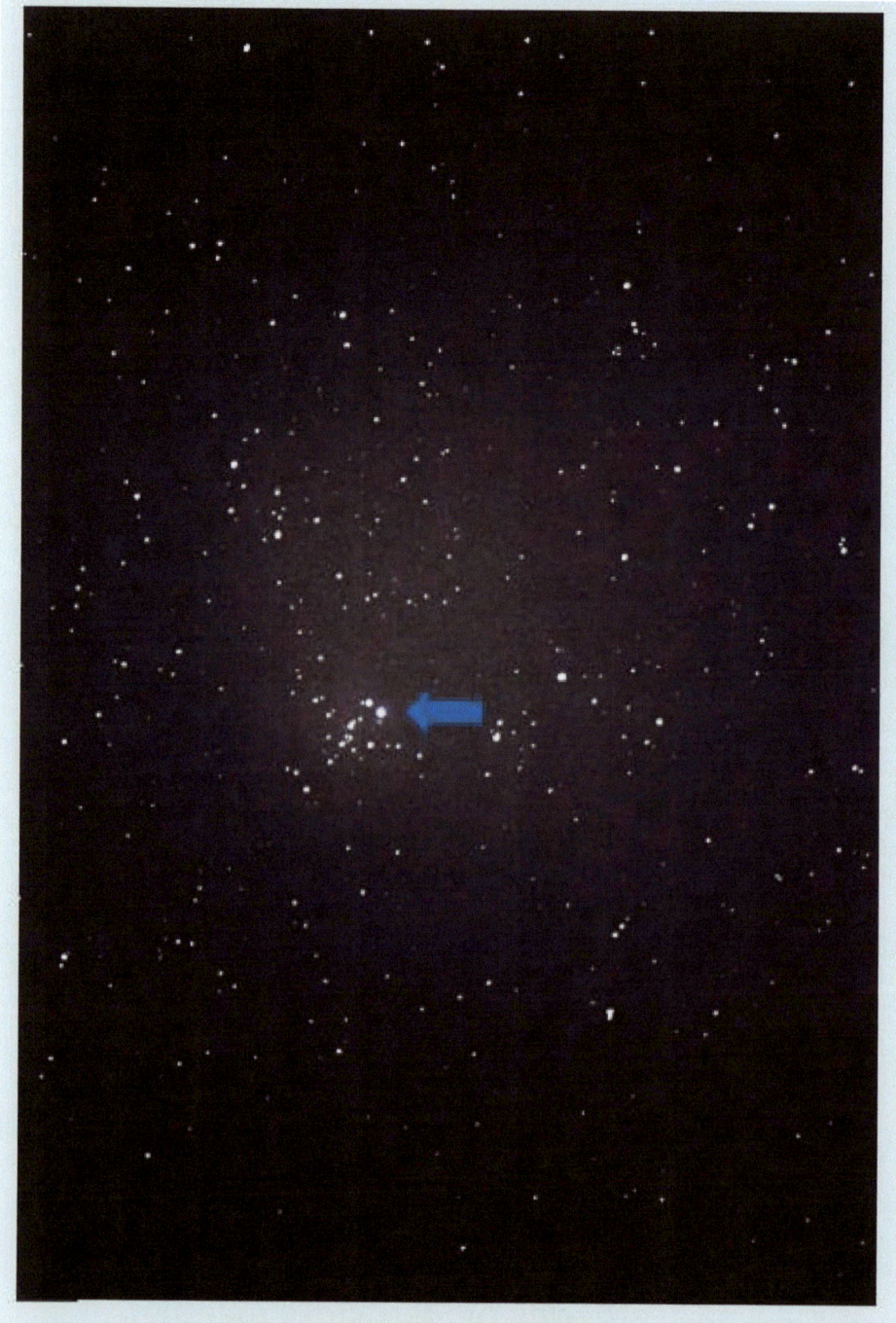

NGC 457

M 50

M 93

M48

M6 Butterfly Cluster

M 7 Ptolemy Cluster

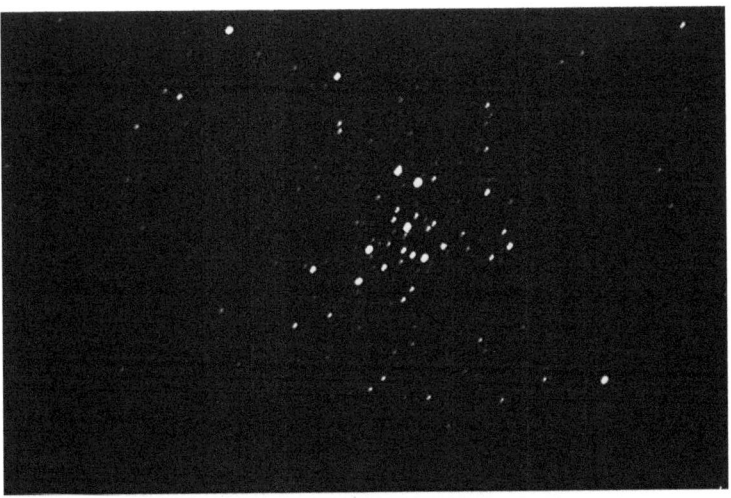

C 76 Jewell Box

The last three pictures of Stars show three Clusters best seen in the Summer Months or early Morning during Winter. The Butterfly Cluster is right in the middle between the Constellation Sagittarius and Scorpius in the direction of the center of the Galaxy. Ptolemy Cluster is more to the South and is an easy target for any Binocular and Telescope. The Jewel Box located at the Southern Portion of the Constellation Scorpius is composed of quite a few number of Stars packed in a small region and because there is a lot of Cosmic Dust in its line of sight the Stars appear to twinkle thus the reason for its name.

M 38 at the tip of blue arrow and NGC 1907 Open Cluster at the top right. These two Clusters are located in the center of Constellation Auriga in a region of intense number of Stars.

M 36 is in the midpoint between M 38 and M 37 adorning the Eastern Portion of Auriga like a frozen firework. Open Clusters are common along the band of the Milky Way Galaxy which is the reason for this alignment between M 38, M36, and M37. The band of the Milky Way is where more Stars are found and New Stars are being born in Nebulas.

M 37 is a very dense Open Cluster that can appear to be a Globular Cluster due to the number Stars viewed in proximity to one another. Around the Cluster there are also many other Stars in a region of intense Star Formation which is right along the Disc of the Milky Way.

Nebulas and Galaxies

Andromeda Galaxy

The closest Galaxy to us and moving in our direction destined to collide with our Galaxy in 4.5 Billion of years from now. Nothing remains the same forever in the Cosmos and collisions between celestial Objects do happen often.

M 8

M 8

Nebula surrounding a Star Cluster possibly formed from the Nebula itself. Located in the Northern Portion of the Constellation Sagittarius and bright even under Light Pollution. It is positioned close to the horizon as seen from the Northern Hemisphere, which is the reason why it is so hard to view it. Proves that many awesome sights might be located under the horizon, but these are only visible from nations farther South. The Earth is a Globe and only parts of the Sky can be seen from certain locations as proof that Earth is not Flat.

Dumbbell Nebula

Difficult to see Cluster at Constellation Vulpecula it appears as a blueish green cloud in Space. It is a Planetary Nebula as a result of the death of a Star similar to our Sun. Vulpecula is a Constellation located in the Center of the Summer Triangle to the North of the Constellation Aquila. It is a great target seen in the Summer Evenings.

Dumbbell Nebula

Orion Nebula

Ring Nebula

Another one of the few Nebulas easily seen from big cities is the Ring Nebula. Located in the Constellation Lyra in a region of great number of Stars along the band of the Milky Way adorning the Summer Nights. Unfortunately, it does not appear very obvious when seen by a human eye, but DSLR Cameras better reveals its shape and colors.

The Ring Nebula is a Planetary Nebula which are the remains after the death of a Sun-like Star. In the Center of the Nebula there is a White Dwarf in its long process of cooling and fading in Space.

Ring Nebula

Ring Nebula

The future of all Planetary Nebulas is their expansion in Space causing their loss of brightness and leading to their fading into interstellar medium. Eventually by the time that the Nebulas currently seen from the Earth are gone, new Nebulas will appear in the sky from the death of other Stars, so it will hardly be impossible to not see any Nebulas at any given point in Time.

Orion Nebula

Orion Nebula

The King of all Nebulas as seen from the Northern Hemisphere is the Orion Nebula. It is a favorite for Astrophotographers, and it can be seen even under heavily light polluted skies of a big city.

The Orion Beula is located in the middle of a Cluster that is slightly South of the Orion Belt. It is possible that the Cluster formed form the Nebula itself which means that in the past the Nebula was bigger and even brighter. These Clouds are the remains of a very large Star that was located not very far from the Sun. In the very center of the Nebula there is a bright Light where there is an active region of Star Formation giving birth to New Stars to spread in the Cosmos. After the Nebula is gone in the future there will be a Star Cluster at its present location.

M 82 Cigar Galaxy seen sideways

M 81 Bodes Galaxy

Bodes Galaxy

Galaxies viewed from the Earth in the past were called Nebulas for their faint appearance. Today we know that these are not Nebulas but rather Island Universes similar to our Milky Way which house Billions of Stars and there are Trillions of Galaxies known now to exist in the Cosmos.

Cigar and Bodes Galaxy.

These are two Galaxies easily spotted near the Constellation Ursa Major.

They seem to be near neighbors of each other.

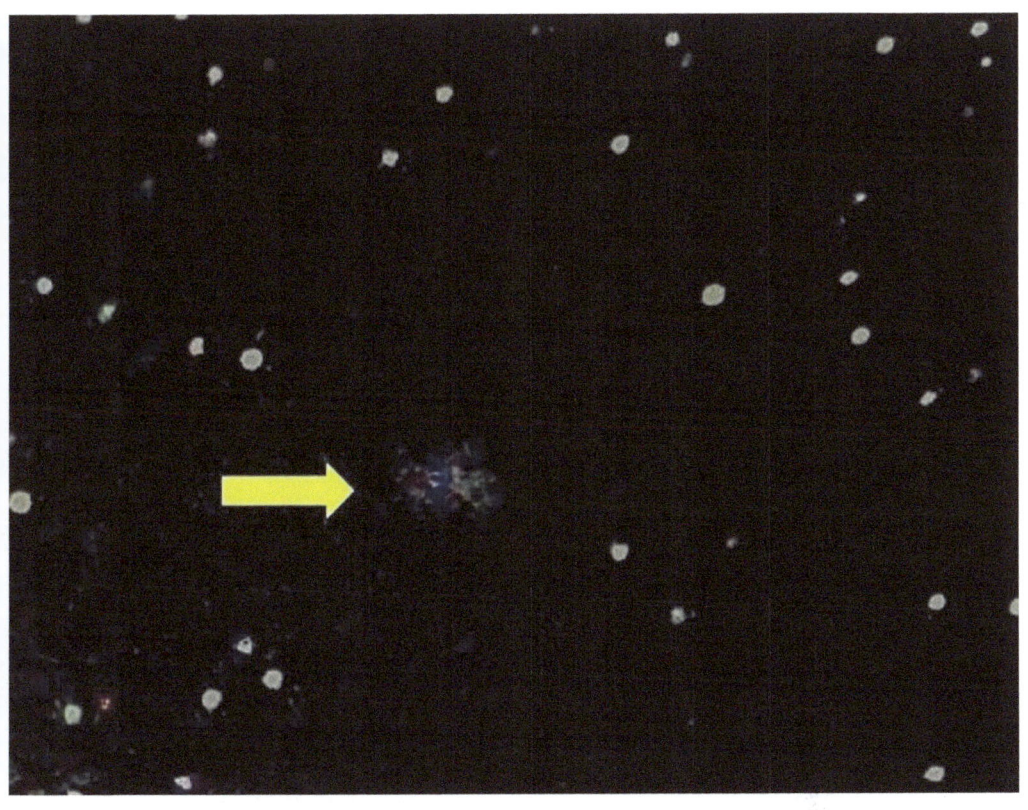

Crab Nebula at tip of arrow shows the aftermath of a recent death of a Star. The Nebula contains a Neutron Star at its Center and was formed after a Supernova that was seen from the Earth in the year 1054 AC. It is located at about 7000 Light Years away and its Neutron Star at its center is a Pulsar which spins many times a second emitting light at several Frequencies form Gamma Rays to Radio Waves. Finding this Nebula in the sky is difficult since it is at an Apparent Magnitude of 8.0 which is very faint for Light Polluted Skies.

Sombrero Galaxy at the tip of the arrow. A very bright Galaxy seen sideways in the Constellation Virgo.

Eclipses

Total Solar Eclipse April 8, 2024, in Fort Worth Texas.

Solar Eclipse Dallas October 10, 2023

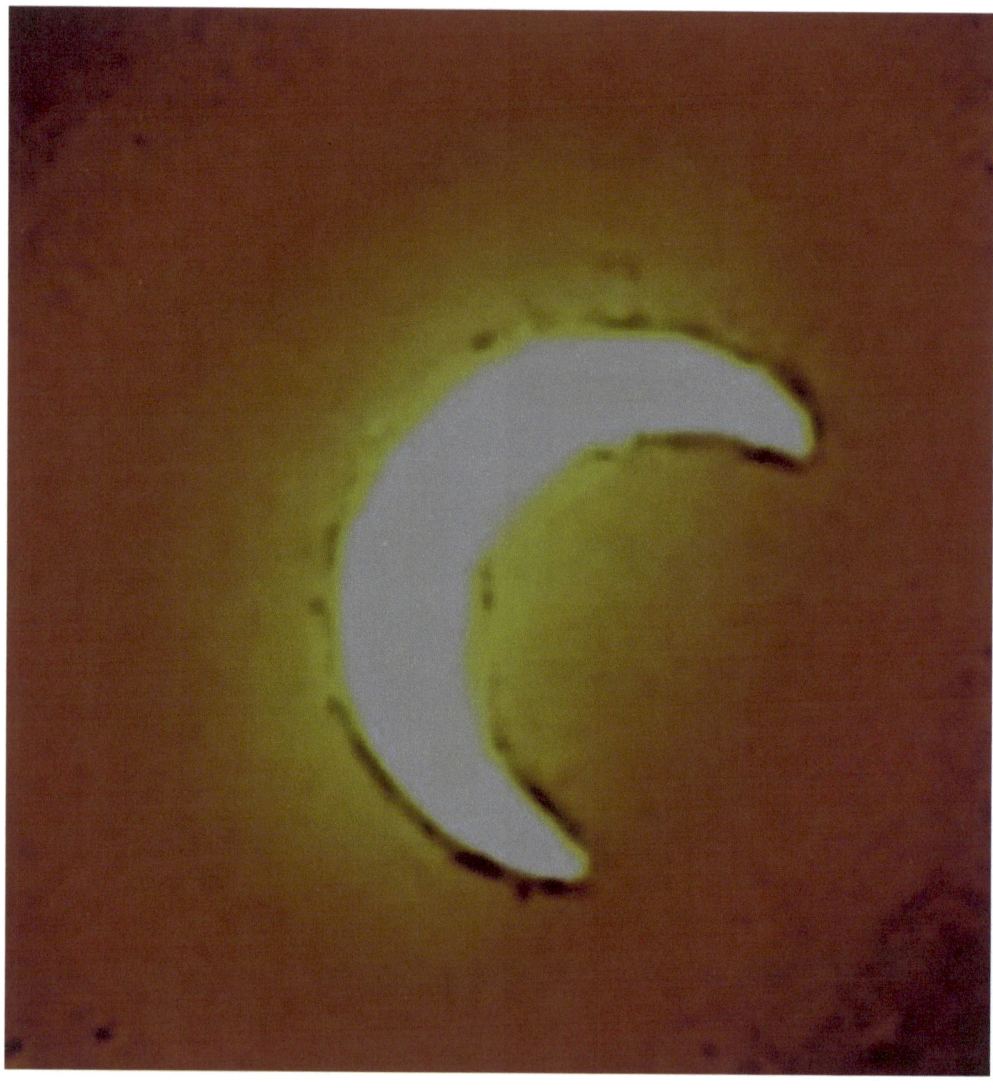

Solar Eclipses happen when the Moon is positioned in front of the Sun as viewed from the Earth. Since the Earth is a Globe, to see the Eclipse depends on the Angle of an Earthly location to have a clear view spot on of the event. Only certain parts of the Earth are in the shadow of the Eclipse where the Moon covers the light from the Sun.

Comet Neowise 2020

This was a Comet that passed by the Earth in 2020 which was a year of many challenges to Humanity, a worldwide Epidemic, and a difficult Political situation. Comets in the past were viewed as bringers of revolutions or major World Disasters. Surprisingly enough, 2020 was a Year different from the others and besides a Comet there was also a Conjunction between Jupiter and Saturn. What a year that was!

In the next few pages there is a table containing 44 targets in the Night Sky within reach of an 8-inch Telescope and a DSLR Camera in the place of an Eyepiece and there are also maps of the Night Sky to help located these Targets. What we can learn from these images is that Space then is filled with wonders and it is a gift to be able to see these Worlds from the Earth. The Ancient Civilizations were able to see the Night Sky better than us in Modern Times, and how much has these sights changed and molded their Cultures. These Celestial Objects were used to Measure the flow of Time and were seen as signs and indications of events on the Earth. Through Astrology came Astronomy and a great understanding of the Night Sky with great many maps of positions of Celestial Objects. The Celestial View and Messier objects change with Time, but in very slow motion when compared to Human Existence. The Stars and Nebulas are not very different today than it was for the Ancients because Space is very large and although Celestial Objects move fast, they appear stationary seen from a far away distance. This makes the sights frozen in Time, and unchanged during a Human Existence. It may take Millions if not Billions of Years for a significant change in the Sky to occur and rarely these changes happen in a Human Lifetime. Slowly the Cosmos changes with Time.

Tables

	Object	Constellation	Story
1	Center of Northern Cross	Cygnus	Multiple Stars.
2	Perseus Double Cluster	Perseus	Binary Cluster.
3	Sagittarius Star Cloud	Sagittarius	Near the direction of the center of Galaxy.
4	Pleiades	Taurus	The Seven Sister Stars.
5	Albireo	Cygnus	Red and Blue Star Binary.
6	C 50	Monoceros	Bright Cluster in the faint Rosette Nebula.
7	Little Beehive	Canis major	Cluster near Star Sirius.

8	C 64	Canis Major	Open Cluster shaped like a shield.
9	Andromeda Galaxy	Andromeda	Brightest Galaxy in the Night Sky.
10	M 8	Sagittarius	Part of the Lagoon Nebula.
11	Dumbbell Nebula	Vulpecula	Blueish Green Nebula in the center of Summer Triangle.
12	Orion Nebula	Orion	Brightest Nebula in the Night Sky.
13	Ring Nebula	Lyra	Nebula shaped like a Ring in Space.

14	Hercules Globular Cluster	Hercules	Most popular Globular Cluster in the Night Sky.
15	Beehive Cluster	Cancer	Open Cluster at the heart of Cancer.
16	M47	Puppis	Open Cluster
17	M46	Puppis	Open Cluster
18	M35	Gemini	Large Open Cluster in the Western portion of Gemini
19	NGC 457	Cassiopeia	Open Cluster shaped like a shield.

20	Cluster 37	Orion	The Cluster is shaped like the numbers 3 and 7.
21	M 50	Monoceros	Open Cluster
22	145 Canis Majoris	Canis Major	Binary Star
23	M 93	Puppis	Open Cluster
24	Cigar Galaxy M 82	Ursa Major	Seen sideways shaped like a cigar.
25	Cor Caroli Double Star	Canes Venatici	Two white Double Star
26	Mizar Double Star	Ursa Major	Double Star
27	M 81 Bodes Galaxy	Ursa Major	Bright light near the Cigar Galaxy
28	M 4 Globular Cluster	Scorpius	Faint and sparse Globular Cluster.

29	M 71 Globular Cluster	Vulpecula	Faint Globular Cluster
30	M 3 Globular Cluster	Canes Venatici	Bright Globular Cluster near the Star Arcturus.
31	M 48	Hydra	Sparse Open Cluster
32	M 12 Globular Cluster	Ophiuchus	Globular Cluster in inside of Ophiuchus.
33	M 6 Butterfly Cluster	Scorpius	Shaped like a butterfly.
34	M 22 Globular Cluster	Sagittarius	Large Globular Cluster.
35	M 7	Scorpius	Stars spread over a large region of the sky.

36	M 10 Globular Cluster	Ophiuchus	Bright Globular Cluster in the center of Ophiuchus.
37	C 76 Jewel Box	Scorpius	Stars of several colors that appear to twinkle.
38	NGC 1907 Open Cluster	Auriga	Broad Cluster slightly South of M 38
39	M 38 Open Cluster	Auriga	Very faint and small cluster
40	M 36 Open Cluster	Auriga	Cluster in the middle point between M 38 and M37

41	M 37 Open Cluster	Auriga	Very bright and dense Open Cluster that resembles a Globular Cluster
42	Crab Nebula	Taurus	Very Faint Nebula of a recent death of a Star.
43	M5	Serpens	Compact Globular Cluster West of Serpens.
44	Sombrero Galaxy	Virgo	Very bright Galaxy seen sideways.

About the Author

I, Diogo Franklin de Souza, was born in the city of Rio de Janeiro, Brazil in August 20, 1986. I moved to Dallas, Texas when I was 11years old. I write stories since I was 9 years old. My books tend to contain short summaries of the most important things I find about life, morality, philosophy, and science. Like I say, everything is part of a whole system, and this is also for everything I do and write. I always wanted to have all the most important knowledge in only a few short books. That is why I write, and that is my inspiration for short summaries. I hope this book brings some inspiration also for the readers, because that really is the purpose of my work. Read it and take from it, pieces of gold for you that can be useful in your life. Enjoy….

www.ingramcontent.com/pod-product-compliance
Lightning Source LLC
Chambersburg PA
CBHW040317220526
45473CB00009B/2467